MW00888215

Made in United States
Troutdale, OR
02/20/2024

17763944R00070